This planner belongs to:

_____

© Copyright 2021 by Happy Oak Tree Press. All rights reserved.

Twenty-two

January
S	M	T	W	T	F	S
						1
2	3	4	5	6	7	8
9	10	11	12	13	14	15
16	17	18	19	20	21	22
23	24	25	26	27	28	29
30	31					

February
S	M	T	W	T	F	S
		1	2	3	4	5
6	7	8	9	10	11	12
13	14	15	16	17	18	19
20	21	22	23	24	25	26
27	28					

March
S	M	T	W	T	F	S
		1	2	3	4	5
6	7	8	9	10	11	12
13	14	15	16	17	18	19
20	21	22	23	24	25	26
27	28	29	30	31		

April
S	M	T	W	T	F	S
					1	2
3	4	5	6	7	8	9
10	11	12	13	14	15	16
17	18	19	20	21	22	23
24	25	26	27	28	29	30

May
S	M	T	W	T	F	S
1	2	3	4	5	6	7
8	9	10	11	12	13	14
15	16	17	18	19	20	21
22	23	24	25	26	27	28
29	30	31				

June
S	M	T	W	T	F	S
			1	2	3	4
5	6	7	8	9	10	11
12	13	14	15	16	17	18
19	20	21	22	23	24	25
26	27	28	29	30		

July
S	M	T	W	T	F	S
					1	2
3	4	5	6	7	8	9
10	11	12	13	14	15	16
17	18	19	20	21	22	23
24	25	26	27	28	29	30
31						

August
S	M	T	W	T	F	S
	1	2	3	4	5	6
7	8	9	10	11	12	13
14	15	16	17	18	19	20
21	22	23	24	25	26	27
28	29	30	31			

September
S	M	T	W	T	F	S
				1	2	3
4	5	6	7	8	9	10
11	12	13	14	15	16	17
18	19	20	21	22	23	24
25	26	27	28	29	30	

October
S	M	T	W	T	F	S
						1
2	3	4	5	6	7	8
9	10	11	12	13	14	15
16	17	18	19	20	21	22
23	24	25	26	27	28	29
30	31					

November
S	M	T	W	T	F	S
		1	2	3	4	5
6	7	8	9	10	11	12
13	14	15	16	17	18	19
20	21	22	23	24	25	26
27	28	29	30			

December
S	M	T	W	T	F	S
				1	2	3
4	5	6	7	8	9	10
11	12	13	14	15	16	17
18	19	20	21	22	23	24
25	26	27	28	29	30	31

Year in Pixels

	J	F	M	A	M	J	J	A	S	O	N	D
1.												
2.												
3.												
4.												
5.												
6.												
7.												
8.												
9.												
10.												
11.												
12.												
13.												
14.												
15.												
16.												
17.												
18.												
19.												
20.												
21.												
22.												
23.												
24.												
25.												
26.												
27.												
28.												
29.												
30.												
31.												

Color Codes

Notes

January 2022

MONDAY	TUESDAY	WEDNESDAY	THURSDAY
3	4	5	6
10	11	12	13
17	18	19	20
24	25	26	27

January 2022

FRIDAY	SATURDAY	SUNDAY	NOTES
	1	2	○
			○
			○
			○
			○
7	8	9	○
			○
			○
			○
14	15	16	○
			○
			○
			○
			○
21	22	23	○
			○
			○
			○
			○
28	29	30	31

February 2022

MONDAY	TUESDAY	WEDNESDAY	THURSDAY
	1	2	3
7	8	9	10
14	15	16	17
21	22	23	24
28			

February 2022

FRIDAY	SATURDAY	SUNDAY	NOTES
4	5	6	○
			○
			○
			○
			○
11	12	13	○
			○
			○
			○
18	19	20	○
			○
			○
			○
			○
25	26	27	○
			○
			○
			○
			○
			NOTES

March 2022

MONDAY	TUESDAY	WEDNESDAY	THURSDAY
	1	2	3
7	8	9	10
14	15	16	17
21	22	23	24
28	29	30	31

March 2022

FRIDAY	SATURDAY	SUNDAY	NOTES
4	5	6	○
			○
			○
			○
			○
11	12	13	○
			○
			○
			○
18	19	20	○
			○
			○
			○
25	26	27	○
			○
			○
			○
			○
			NOTES

April 2022

MONDAY	TUESDAY	WEDNESDAY	THURSDAY
4	5	6	7
11	12	13	14
18	19	20	21
25	26	27	28

April 2022

FRIDAY	SATURDAY	SUNDAY	NOTES
1	2	3	○
			○
			○
			○
			○
8	9	10	○
			○
			○
			○
15	16	17	○
			○
			○
			○
			○
22	23	24	○
			○
			○
			○
			○
29	30	1	NOTES

May 2022

MONDAY	TUESDAY	WEDNESDAY	THURSDAY
2	3	4	5
9	10	11	12
16	17	18	19
23	24	25	26
30	31		

May 2022

FRIDAY	SATURDAY	SUNDAY	NOTES
6	7	8	○
			○
			○
			○
13	14	15	○
			○
			○
			○
20	21	22	○
			○
			○
			○
			○
27	28	29	○
			○
			○
			○
			○
			NOTES

June 2022

MONDAY	TUESDAY	WEDNESDAY	THURSDAY
		1	2
6	7	8	9
13	14	15	16
20	21	22	23
27	28	29	30

June 2022

FRIDAY	SATURDAY	SUNDAY	NOTES
3	4	5	○
			○
			○
			○
			○
10	11	12	○
			○
			○
			○
17	18	19	○
			○
			○
			○
			○
24	25	26	○
			○
			○
			○
			○
			NOTES

July 2022

MONDAY	TUESDAY	WEDNESDAY	THURSDAY
4	5	6	7
11	12	13	14
18	19	20	21
25	26	27	28

July 2022

FRIDAY	SATURDAY	SUNDAY	NOTES
1	2	3	○
			○
			○
			○
			○
8	9	10	○
			○
			○
			○
15	16	17	○
			○
			○
			○
			○
22	23	24	○
			○
			○
			○
			○
29	30	31	NOTES

August 2022

MONDAY	TUESDAY	WEDNESDAY	THURSDAY
1	2	3	4
8	9	10	11
15	16	17	18
22	23	24	25
29	30	31	

August 2022

FRIDAY	SATURDAY	SUNDAY	NOTES
5	6	7	○
			○
			○
			○
			○
12	13	14	○
			○
			○
			○
19	20	21	○
			○
			○
			○
			○
26	27	28	NOTES

September 2022

MONDAY	TUESDAY	WEDNESDAY	THURSDAY
			1
5	6	7	8
12	13	14	15
19	20	21	22
26	27	28	29

September 2022

FRIDAY	SATURDAY	SUNDAY	NOTES
2	3	4	○
9	10	11	○
16	17	18	○
23	24	25	○
30			NOTES

October 2022

MONDAY	TUESDAY	WEDNESDAY	THURSDAY
3	4	5	6
10	11	12	13
17	18	19	20
24	25	26	27

October 2022

FRIDAY	SATURDAY	SUNDAY	NOTES
	1	2	○ ○ ○ ○ ○
7	8	9	○ ○ ○ ○
14	15	16	○ ○ ○ ○ ○
21	22	23	○ ○ ○ ○ ○
28	29	30 / 31 MONDAY	NOTES

November 2022

MONDAY	TUESDAY	WEDNESDAY	THURSDAY
	1	2	3
7	8	9	10
14	15	16	17
21	22	23	24
28	29	30	

November 2022

FRIDAY	SATURDAY	SUNDAY	NOTES
4	5	6	○
			○
			○
			○
			○
11	12	13	○
			○
			○
			○
18	19	20	○
			○
			○
			○
			○
25	26	27	○
			○
			○
			○
			○
			NOTES

December 2022

MONDAY	TUESDAY	WEDNESDAY	THURSDAY
			1
5	6	7	8
12	13	14	15
19	20	21	22
26	27	28	29

December 2022

FRIDAY	SATURDAY	SUNDAY	NOTES
2	3	4	○
9	10	11	○
16	17	18	○
23	24	25	○
30	31		NOTES

December
2021

01 WEDNESDAY

02 THURSDAY

03 FRIDAY

04 SATURDAY

December
2021

05 SUNDAY

06 MONDAY

07 TUESDAY

08 WEDNESDAY

09 THURSDAY

10 FRIDAY

11 SATURDAY

12 SUNDAY

December 2021

13 MONDAY

14 TUESDAY

15 WEDNESDAY

16 THURSDAY

December 2021

17 FRIDAY

18 SATURDAY

19 SUNDAY

20 MONDAY

December
2021

21 TUESDAY

22 WEDNESDAY

23 THURSDAY

24 FRIDAY

December 2021

25 SATURDAY

26 SUNDAY

27 MONDAY

28 TUESDAY

December 2021

29 WEDNESDAY

30 THURSDAY

31 FRIDAY

NOTES

January 2022

01 SATURDAY

02 SUNDAY

03 MONDAY

04 TUESDAY

January 2022

05 WEDNESDAY

06 THURSDAY

07 FRIDAY

08 SATURDAY

January 2022

09 SUNDAY

10 MONDAY

11 TUESDAY

12 WEDNESDAY

January 2022

13 THURSDAY

14 FRIDAY

15 SATURDAY

16 SUNDAY

January 2022

17 MONDAY

18 TUESDAY

19 WEDNESDAY

20 THURSDAY

January 2022

21 FRIDAY

22 SATURDAY

23 SUNDAY

24 MONDAY

January 2022

25 TUESDAY

26 WEDNESDAY

27 THURSDAY

28 FRIDAY

January 2022

29 SATURDAY

30 SUNDAY

31 MONDAY

NOTES

01 TUESDAY

02 WEDNESDAY

03 THURSDAY

04 FRIDAY

05 SATURDAY

06 SUNDAY

07 MONDAY

08 TUESDAY

09 WEDNESDAY

10 THURSDAY

11 FRIDAY

12 SATURDAY

13 SUNDAY

14 MONDAY

15 TUESDAY

16 WEDNESDAY

17 THURSDAY

18 FRIDAY

19 SATURDAY

20 SUNDAY

21 MONDAY

22 TUESDAY

23 WEDNESDAY

24 THURSDAY

25 FRIDAY

26 SATURDAY

27 SUNDAY

28 MONDAY

March 2022

01 TUESDAY

02 WEDNESDAY

03 THURSDAY

04 FRIDAY

March 2022

05 SATURDAY

06 SUNDAY

07 MONDAY

08 TUESDAY

09 WEDNESDAY

10 THURSDAY

11 FRIDAY

12 SATURDAY

March 2022

13 SUNDAY

14 MONDAY

15 TUESDAY

16 WEDNESDAY

March 2022

17 THURSDAY

18 FRIDAY

19 SATURDAY

20 SUNDAY

21 MONDAY

22 TUESDAY

23 WEDNESDAY

24 THURSDAY

March 2022

25 FRIDAY

26 SATURDAY

27 SUNDAY

28 MONDAY

29 TUESDAY

30 WEDNESDAY

31 THURSDAY

NOTES

April 2022

01 FRIDAY

02 SATURDAY

03 SUNDAY

04 MONDAY

05 TUESDAY

06 WEDNESDAY

07 THURSDAY

08 FRIDAY

April 2022

09 SATURDAY

10 SUNDAY

11 MONDAY

12 TUESDAY

13 WEDNESDAY

14 THURSDAY

15 FRIDAY

16 SATURDAY

April 2022

17 SUNDAY

18 MONDAY

19 TUESDAY

20 WEDNESDAY

21 THURSDAY

22 FRIDAY

23 SATURDAY

24 SUNDAY

April 2022

25 MONDAY

26 TUESDAY

27 WEDNESDAY

28 THURSDAY

April 2022

29 FRIDAY
-
-
-
-
-
-
-
-

30 SATURDAY
-
-
-
-
-
-
-
-

NOTES

May 2022

01 SUNDAY

02 MONDAY

03 TUESDAY

04 WEDNESDAY

May 2022

05 THURSDAY

06 FRIDAY

07 SATURDAY

08 SUNDAY

May 2022

09 MONDAY

10 TUESDAY

11 WEDNESDAY

12 THURSDAY

May 2022

13 FRIDAY

14 SATURDAY

15 SUNDAY

16 MONDAY

May
2022

17 TUESDAY

18 WEDNESDAY

19 THURSDAY

20 FRIDAY

May 2022

21 SATURDAY

22 SUNDAY

23 MONDAY

24 TUESDAY

May 2022

25 WEDNESDAY

26 THURSDAY

27 FRIDAY

28 SATURDAY

May 2022

29 SUNDAY

30 MONDAY

31 TUESDAY

NOTES

June 2022

01 WEDNESDAY

02 THURSDAY

03 FRIDAY

04 SATURDAY

June 2022

05 SUNDAY

06 MONDAY

07 TUESDAY

08 WEDNESDAY

June 2022

09 THURSDAY

10 FRIDAY

11 SATURDAY

12 SUNDAY

June 2022

13 MONDAY

14 TUESDAY

15 WEDNESDAY

16 THURSDAY

June 2022

17 FRIDAY

18 SATURDAY

19 SUNDAY

20 MONDAY

June 2022

21 TUESDAY

22 WEDNESDAY

23 THURSDAY

24 FRIDAY

June 2022

25 SATURDAY

26 SUNDAY

27 MONDAY

28 TUESDAY

June 2022

29 WEDNESDAY

30 THURSDAY

NOTES

July 2022

01 FRIDAY

02 SATURDAY

03 SUNDAY

04 MONDAY

July 2022

05 TUESDAY

06 WEDNESDAY

07 THURSDAY

08 FRIDAY

July 2022

09 SATURDAY

10 SUNDAY

11 MONDAY

12 TUESDAY

July 2022

13 WEDNESDAY

14 THURSDAY

15 FRIDAY

16 SATURDAY

July 2022

17 SUNDAY

18 MONDAY

19 TUESDAY

20 WEDNESDAY

July 2022

21 THURSDAY

22 FRIDAY

23 SATURDAY

24 SUNDAY

July 2022

25 MONDAY

26 TUESDAY

27 WEDNESDAY

28 THURSDAY

July 2022

29 FRIDAY

30 SATURDAY

31 SUNDAY

NOTES

August 2022

01 MONDAY
-
-
-
-
-
-
-
-

02 TUESDAY
-
-
-
-
-
-
-
-
-

03 WEDNESDAY
-
-
-
-
-
-
-
-

04 THURSDAY
-
-
-
-
-

05 FRIDAY

06 SATURDAY

07 SUNDAY

08 MONDAY

August 2022

09 TUESDAY

10 WEDNESDAY

11 THURSDAY

12 FRIDAY

August 2022

13 SATURDAY

14 SUNDAY

15 MONDAY

16 TUESDAY

August 2022

17 WEDNESDAY

18 THURSDAY

19 FRIDAY

20 SATURDAY

August 2022

21 SUNDAY

22 MONDAY

23 TUESDAY

24 WEDNESDAY

August 2022

25 THURSDAY

26 FRIDAY

27 SATURDAY

28 SUNDAY

29 MONDAY

30 TUESDAY

31 WEDNESDAY

NOTES

September 2022

01 THURSDAY

02 FRIDAY

03 SATURDAY

04 SUNDAY

September 2022

05 MONDAY

06 TUESDAY

07 WEDNESDAY

08 THURSDAY

September 2022

09 FRIDAY

10 SATURDAY

11 SUNDAY

12 MONDAY

September 2022

13 TUESDAY

14 WEDNESDAY

15 THURSDAY

16 FRIDAY

September 2022

17 SATURDAY

18 SUNDAY

19 MONDAY

20 TUESDAY

September 2022

21 WEDNESDAY

22 THURSDAY

23 FRIDAY

24 SATURDAY

September 2022

25 SUNDAY

26 MONDAY

27 TUESDAY

28 WEDNESDAY

September 2022

29 THURSDAY

30 FRIDAY

NOTES

01 SATURDAY

02 SUNDAY

03 MONDAY

04 TUESDAY

October 2022

05 WEDNESDAY

-
-
-
-
-
-
-
-

06 THURSDAY

-
-
-
-
-
-
-
-

07 FRIDAY

-
-
-
-
-
-
-
-

08 SATURDAY

-
-
-
-
-

October 2022

09 SUNDAY

10 MONDAY

11 TUESDAY

12 WEDNESDAY

October 2022

13 THURSDAY

14 FRIDAY

15 SATURDAY

16 SUNDAY

October
2022

17 MONDAY

18 TUESDAY

19 WEDNESDAY

20 THURSDAY

October 2022

21 FRIDAY

22 SATURDAY

23 SUNDAY

24 MONDAY

October 2022

25 TUESDAY

26 WEDNESDAY

27 THURSDAY

28 FRIDAY

October 2022

29 SATURDAY

30 SUNDAY

31 MONDAY

NOTES

November 2022

01 TUESDAY

02 WEDNESDAY

03 THURSDAY

04 FRIDAY

November 2022

05 SATURDAY

06 SUNDAY

07 MONDAY

08 TUESDAY

November 2022

09 WEDNESDAY

10 THURSDAY

11 FRIDAY

12 SATURDAY

November 2022

13 SUNDAY

14 MONDAY

15 TUESDAY

16 WEDNESDAY

November 2022

17 THURSDAY

18 FRIDAY

19 SATURDAY

20 SUNDAY

21 MONDAY

22 TUESDAY

23 WEDNESDAY

24 THURSDAY

November 2022

25 FRIDAY

26 SATURDAY

27 SUNDAY

28 MONDAY

November 2022

29 TUESDAY

-
-
-
-
-
-
-
-

30 WEDNESDAY

-
-
-
-
-
-
-
-

NOTES

December 2022

01 THURSDAY

02 FRIDAY

03 SATURDAY

04 SUNDAY

December 2022

05 MONDAY

06 TUESDAY

07 WEDNESDAY

08 THURSDAY

09 FRIDAY

10 SATURDAY

11 SUNDAY

12 MONDAY

December 2022

13 TUESDAY

14 WEDNESDAY

15 THURSDAY

16 FRIDAY

December 2022

17 SATURDAY

18 SUNDAY

19 MONDAY

20 TUESDAY

December 2022

21 WEDNESDAY

22 THURSDAY

23 FRIDAY

24 SATURDAY

December 2022

25 SUNDAY

26 MONDAY

27 TUESDAY

28 WEDNESDAY

December 2022

29 THURSDAY

30 FRIDAY

31 SATURDAY

NOTES

www.ingramcontent.com/pod-product-compliance
Lightning Source LLC
Chambersburg PA
CBHW081310070526
44578CB00006B/824